THE COMPASS CHRONICLES

HOW TO FIND YOUR WAY AROUND SPACE IN JUST TWO TIMES

DANIEL A. MARTENS
YAVERBAUM

Executive Editor HANNAH H. CHU

Editor MARSDEN EPWORTH

Made Possible by THE LAKEVILLE JOURNAL
& TRICORNER NEWS

Tattenwitch Editions, being an imprint of
LAW FOUR PRESS
Brooklyn, NY

〜

ATTRIBUTION & ACKNOWLEDGMENT

Neither this collection nor any of the collected items would have been possible without the opportunity, enthusiasm and permission provided by the Lakeville Journal editorial staffs of 2007 through 2018. Without Marsden Epworth in particular, it's hard to imagine that any of the columns to appear here would have warranted the original newspaper space; they most certainly would not have been submitted in time.

To the best of our understanding, every reproduction included herein is of an image belonging to the public domain.

A number of the illustrations used in this text, however, were made expressly for such purpose. With one exception, every original graphic was designed, produced and coordinated by Hannah Chu. Graphic management turned out to constitute but one of many ways in which this project's completion was secured at Hannah's hand.

Neither artistic innovation nor scientific precision, however, was the aim of any picture. The images are included for purposes of instruction. Each picture is intended to support its surrounding paragraphs by providing a means by which to reduce the number of possible objects under discussion, rather than by enhancing the discussion.

Permeating the details and descriptions to follow is, to be sure, an unabashed conviction that the night sky richly rewards any and all efforts of the eye — even or especially the eye unaided by equipment. Such enthusiasm for long and direct gazes at vastness tend eventually to leave little interest in the use of ink and a thumbnail to capture optics — particularly in the case of a black and white text.

The illustration's clarifying and instructing purpose was deemed most directly and least distractingly served, therefore, if computer-rendered. Every figure marked finally for inclusion was generated by means either of the *Starry Night Pro 8* software package (copyright © Simulation Curriculum Corp. 2018, https://starrynighteducation.com, see *Part III, Epilogue,* of this text) or of the *Stellarium* 0.18.2 software package (open source, https://stellarium.org, see *Part III, Epilogue,* of this text).

The author and editor are extremely grateful for the availability of the above described programs.

The single graphic exception to all the above is that on this book's cover. The cover image consists of a stick-figure person sitting contemplatively under a tree. This vignette, aided largely by the then standard software *Microsoft Paint*, was drawn by Jesse Anselm Yaverbaum at the approximate age of ten.

Superimposed on this tree of modernity is a somewhat fanciful swarm of stars. These were sketched by Galileo in approximately 1626 A.D. and meant for inclusion in his work *Sidereus Nuncius* (*The Starry Messenger*).

CONTENTS

PART I

THE AUTUMN AND THE WINTER

PREFACE: CLOCKWISE

THE WHEN AND THE WHEREFORE

*I*rrelevantly long ago, via some geologic processes that might be hard to imagine or appreciate, but are apparently even harder to deny, it seems that a southern swath of what are today Vermont's Green Mountains met a river or vice versa. And the Mountains knew the river. And so the Mountains and the river begot some hill country. And the hill country, which now snakes along a sector of Earth's surface we have come to regard as the western border of Massachusetts, plowed a corridor of comparable scale along contemporary Connecticut's occidental edge. The hill country thereby conferred on two neighboring states a single piece of signature landscape. And all saw that this landscape was good. Very very good.

Find yourself in nearly any sub-region of the fertile candescent known to tourists and taxpayers as *The Berkshires* and you are bound to find: Altitude matters. The amount you've ascended can build in smooth stealth behind your back; the floor you hope to find in front can take a treacherously quick way home — and thereby seem to smuggle all hints of horizontality with it.

Yet starting somewhere above every steadfast Berkshire summit is a height that's so much higher, we might as well call it deep: When willing to push past the peaks to which timberline vectors seem most naturally to guide our gaze, we find ourselves staring at heights of quite a different order. This fresh new scale makes milli-mincemeat of a mile; it leaves our yardsticks supine and squashes out of the tallest terrestrial tower any pretense toward relevance.

Up at the above mentioned above, hangs a stately sweep of transition between atmosphere and space known as sky. A well setting sun can deliver a stunning visual performance, but it is rarely a grand finale: At dusk, we extinguish the house lights in a most vibrant of theaters for among the most long-running productions on Milky Way. Enter stage left: the show of the stars.

Consider, for example, the autumn of 2007 and subsequent winter of 2008. Consider twenty or so evenings distributed more or less evenly throughout the six month span. Now give a serious look into the thought of seriously looking up. See about staring patiently and carefully — enthusiastically but not necessarily knowledgeably — up at the Berkshire sky from, for example, Sheffield, MA[1].

Assume, moreover, some straightforward and standard use of the term '*interesting*': '*noteworthy*', for example, or '*unanticipated*', perhaps; maybe even consider connotations such as *'curious'* or *'anomalous'*, etc.

Given more or less the above described slices of space, time and semantics, the conclusion seems straightforward: Nothing interesting whatsoever occurred. There was simply nothing to see, other than a backdrop in waiting. There was, that is, just whatever simple and default combinations of light and dark in which we might be prepared to believe without seeing.

Each celestial happening was essentially typical, and certainly expected. To the extent that the delivery of drama correlates with divergence from expectation, the sky — at almost all times — just could not be duller. To the extent, however, that we allow for flavors of exhilaration and spectacle to be hewn from satisfaction rather than surprise (symphony crescendos, say, in place of jazz solos), the night sky show is without compare.

On approximately a bi-weekly basis, therefore, in a sky-watch column published in the *Compass* weekend magazine of newspapers across the Berkshires, a libretto was provided. Strung together, the set of naked-eye star-gazing pieces form a sequence of introductory lessons as to where and when and why to look up when so moved.

The casual curriculum applies with no less conceptual or qualitative accuracy to the stars of this year than it did in 2008. For quantitative precision, planet times and places must be adjusted. For a sense of celestial polyrhythms and a taste of cosmic connect-the-dots, however, simply follow the old news to follow.

---

1. Zip code 01257, latitude 42.1° north of Earth's *equator*, longitude 73.3°degrees west of its *prime meridian*

EQUINOX

2007. SEPTEMBER 20

This Sunday, September 23, marks the astronomical beginning of fall: the autumnal equinox. Although an equinox is not an especially visual event, its semi-annual arrival celebrates some of the symmetries and relationships at the heart of astronomy's appeal.

The deep sky is where time and space comingle: the position of the sun is a tick on our clock, the shape of the moon is a page on our calendar. We embrace this delightful confusion at the start of both autumn and spring: An equinox is, by definition, both a particular point in time and a particular point in space.

At 5:51 am (Eastern Daylight Time) on Sunday, the rays of our sun will slam directly and perpendicularly on Earth's equator. If you were to visit the equator and lie face-up on the beach at noon this Sunday, you would find the sun all the way at the tippy- top (*'zenith'*) of the sky. Its beams would shine straight down on your forehead. For all other latitudes on Earth, the sun will be lower in the sky and its rays will graze rather than attack Earth's surface. For us here at approximately 42 degrees above the equator—as for our counterparts at 42 degrees below—we should not lie down to find the sun at noon. We should angle our heads up no higher than 90 – 42 degrees—that is, 48 degrees above the horizontal.

The significance of the equinox is that Earth's northern hemisphere gets no more direct (perpendicular) sunlight than does the southern hemisphere. Approximately, then, the sun will rise earlier than noon by the same amount that it sets later than noon, producing a twenty four hour period equally divided between day and night.

(*'Equinox'* etymologically refers to *'equal night'*.) A variety of reasons, culturally significant but religiously confusing the human imposition of time zones and daylight saving time, the duration of our *equal night* never turns out to not, precisely work out precisely to that of the adjacent *equal ∂ay*.

On Sunday, the first day of autumn, the sun will rise at 6:41 am (20 minutes before what would be the 'perfect' time if we hadn't set our clocks ahead back in April) and set at 6:51 pm (10 minutes before "perfection"). Each of these events will occur singularly near *∂ue* east and *∂ue* west, respectively.

Sunrise & Sunset (top to bottom):
Equinox vs. Solstice (left to right)

*F*or sunlight to hit us in the above fashion, the sun itself must aim from a specific location. This position, currently found in the western portion of the constellation *'Virgo'*, is a place literally called the *'autumnal equinox'*. So, as it all comes together and our sun passes the mid-straightaway of another graceful lap, spend an instant celebrating: Treat your clock and compass to some cosmic scale re-calibrating.

AUTUMN A.M.

OCTOBER 4

We recently received a letter and lovely sketch from a Sheffield resident, *Alfred R.* Alfred and his wife have been enjoying pre-dawn walks. Clear views of Orion looming in the south motivated a question or two. For this coming week in particular, the R's really have the right idea. Much of early autumn's astronomical action will be delivered in the crisp mornings.

First off, the moon now wanes (appears to shrink). It hit its *'last quarter'* phase (which means it looks like the left half of a circle) Wednesday, October 3; it will be new (entirely dark) again next Thursday,

October 11. Whenever the moon experiences the waning half of its monthly cycle, you can expect to find it in the morning sky rather than at night.

Second, some of the more dazzling planets now gather in the mornings. Venus has shifted to its identity as the *'morning star'*. In ancient Athens, this 'bringer of light' was known as *Phosphorus*, while the brightest 'evening star' was called *Hesperus*. Eventually, Greek astronomers accepted the Babylonian view that the two were in fact one: one object that wandered back and forth across the otherwise fixed canopy of stars. The Greek word for 'wanderer' is *planet*. Hence, the club was formed: Venus, Jupiter, Saturn, Mercury, Mars, Sun and Moon were the seven extraordinary creatures that had somehow cut their tethers to the thousands of dots comprising immutable constellations. (Consider the days of the week. In particular, consider the days' names in Spanish, French or the like.) Since the development of telescopes, the rules for membership in this club have gotten a bit more complicated. Just ask Pluto how he feels about the details.

So, you can expect to see Venus rising elegantly in the southeast around 6 am this week. Slightly after Venus rises, Saturn does the same. If you had to pick one day to check out these lustrous wanderers, wake up early Sunday, October 7. Venus, the moon and Saturn will all hang out together—in a tidy line from top right to bottom left. If you notice a fourth (less intense but still quite prominent) object above the moon and off the diagonal line, then you have met Regulus—the brightest star in the constellation Leo.

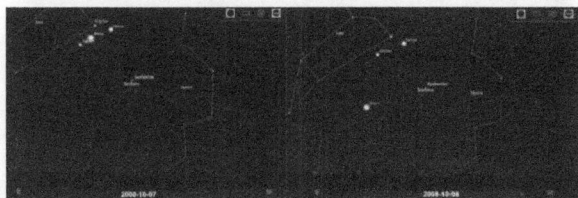

Regulus is comparatively small (about 3 ½ times the mass of our sun) and comparatively young (a few hundred million years old), but it's bright because it's comparatively close: Light only takes about 77 years to travel from Regulus to Earth. In the context of outer space, that's just a photon and his wife taking an early-morning stroll.

CAPRICORN

OCTOBER 11

*A*t approximately 9 pm this week, stare up and due south. For a slight change in pace, look neither for the brightest nor the largest star groupings. Simply look for the most centrally, conveniently, southerly located 'triangle' of stars you can find. This triangle should be high enough to be unblocked by hills, buildings, etc., but low enough to prevent whiplash. If you just let your eyes relax and fall toward the most accessible middle chunk of the southern sky, a seemingly three-sided figure will eventually take shape. The top right corner will be the brightest; the top left corner will be almost as bright; the bottom will be distressingly faint.

Although the last passenger in the world to board your train of thought might be a *sea-goat*, the triangle at which you gaze is nonetheless *Capricorn*: the unassuming protagonist of autumn's zodiac show. Capricorn, will occupy center stage all week, so, despite its comparatively modest appearance, it's a helpful constellation to know. It is one of the oldest to be identified.

On an exceptionally clear night, fix on the stars comprising the triangle and then consider an alternate interpretation: Eliminate the triangle sides from your mind and instead connect the stars with lines that zig-zag from the top left down and up to the top right. With a little finesse and creativity, you may start to relate to the 'goat' evidently lurking (with his bowed head at the left and his luminous tail at the right) behind the dots.

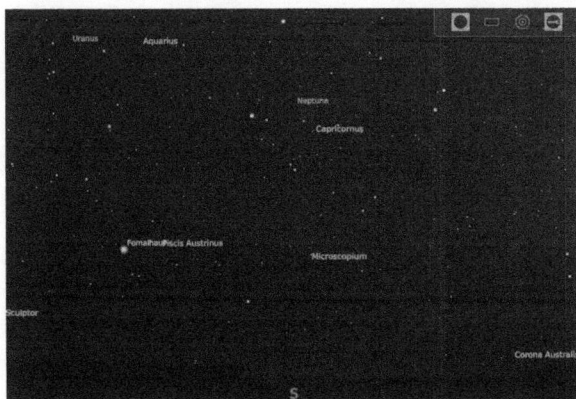

Why a 'sea' goat? Some of Capricorn's closest neighbors are *Aquarius* (the 'Water-Bearer'), *Pisces* (the 'Fish') and *Delphina* (the 'Dolphin'). As goatish as he may appear (albeit after some effort), he lives in a region of the sky classically known to Northern Hemisphere dwellers as the 'sea'. This sky zone shares a deep-down, mysterious, dark nature with underwater worlds in the following sense: When our days are shortest and our light the most stingy, it is down here in the sea of the sky where the sun dwells—hibernates, if you will.

Just as the *equinox*, a few weeks ago, was a point both in time and in space, the *'winter solstice'* refers to something more than the least sunny day of the year. It is the point in space, during antiquity found at Capricorn's tail, from which the sun shines on that day. From there, the rays are aimed as south of us as they ever are: at the *Tropic of Capricorn*.

But why, then, is my friend's birthday in October but she is not a Capricorn? Put another way, why is Capricorn prominent *now* if we have only just begun autumn? Good question. A constellation is clearly visible in the night sky precisely when the sun is *not* located there. If you look up (but do not stare) at the sun during the day some time this

week, then you're looking toward Virgo (and had you looked during antiquity, you would have been looking at Libra—as 'expected'). Its stars are certainly out and up there; they're just nowhere near close enough to compete with the sun's intensity.

METEOR FALL

OCTOBER 18

*D*ig out a see-through umbrella: The outer-space forecast calls for meteor showers. Sunday morning, October 21, the *Orionids* will reach their peak.

If you look in the right direction, you might well observe a 'shooting-star' at any time this month, but the bulk of the storm will be found high in the south-east at approximately 4 am Sunday. In a good shower, a celestial firecracker will zip by roughly every three minutes, yet each one will still somehow feel like an unexpected thrill. As random as the viewing experience might seem, the shooting stars of any shower are all 'shot' from a particular center-point—different for each meteor shower—known as the *radiant*.

For the Orionids, the radiant is in the north-east (top left) corner of the constellation *Orion*—right at its border with the constellation Gemini.

Orion is a singularly lovely constellation. Other than the 'Big Dipper' (which, in fact, is only part of the constellation 'Big Bear'), Orion is quite likely the most recognizable of all the constellations. First of all, Orion is *huge*. When Orion the Hunter is out hunting, you cannot help but know it. He dominates that part of the southern sky which is just high enough to spite buildings even in New York City (well, all right, Brooklyn).

Second, a number of its constituent stars—Betelgeuse, the hunter's left shoulder; Rigel, his right knee—are unusually bright.

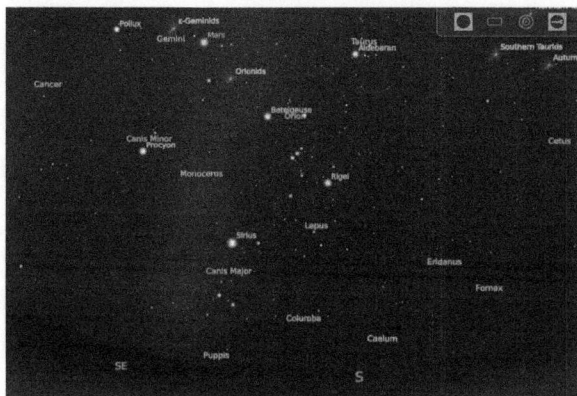

Third, Orion truly does resemble a figure. His 'belt' (composed of three bright stars in a line) is the signature. Once you see it, you can soon visualize a shield held out at right and an up-stretched club-wielding arm to the left. Fourth, Orion is located near the *celestial equator* — a band around the sky that you would find if you took Earth's equator and stretched it away from our surface and into space. This means that Orion the Hunter, unlike many constellations, is familiar to dwellers of both the northern and southern hemispheres of Earth. Deservedly, he's a popular guy and his is a popular spot.

So look up toward Orion this Sunday

morning. The moon will be a waxing gibbous, so it will have set by 1:30 am. If your view is not obstructed by clouds or architecture, you are guaranteed to enjoy the show: On clear nights, meteor showers return an outrageously optical bang for precisely one buck (you do have to wake up). You will be pleased to have met Orion because he and his trusty dog *Sirius* plan to tower over the sky during civilized hours this coming winter.

THE DARK SIDE OF THE MAN
IN THE MOON

OCTOBER 25

*T*his Friday, October 26, the moon will be full. It will be on the side of the Earth opposite from that of the sun. All sun rays that rebound off one hemisphere of the moon will find us in their path. Actually, the moon will pass through the precise point of complete *opposition* at 12:52 am Friday, so it will look just about as full Thursday night as it will Friday night.

On an autumn night when the moon is full, it is generally the place to look. First, the full moon is a visual gift that keeps on giving (approximately twelve times a year); second, the moon's light dominates the sky, impeding vivid naked-eye observation of background or nearby constellations. Friday night, for example, the moon will be found in that part of the sky known as *Aries*, which means that the stars of *Pisces*, Aries and *Taurus* will yield the floor to the great but graceful lunar disc.

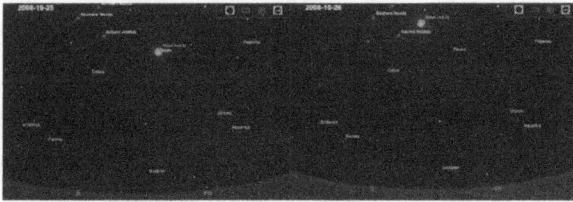

When winter comes, a number of Taurus's sublime constituents and companions will grow familiar, Aldebaran and the 'seven sisters' of the *Pleiades*, to name two. But this week, make eye contact with the 'man in the moon'. You know that face will be there—staring at you from the same angle it always does. No matter what month or year, our sole natural satellite never turns its back on us.

We Earthlings always gaze at approximately

the same eerily humanoid map— chiefly characterized by a seeming left eye called *Mare Imbrium* ('Sea of Rain'), a seeming right eye called *Mare Serenitatis* ('Sea of Serenity') and a mouth called *Mare Nubium* ('Sea of Clouds'). These *lunar maria* are not actual water bodies, but large, low, flat and iron-rich plains formed by ancient volcanic eruptions. Their depth and iron content produce pockets of darkness with which we grow almost unconsciously familiar.

Certainly, the moon possesses a 'back' and this back often gets illuminated by the sun. No extra-terrestrial neighbor of ours (imagined or discovered) would understand what we meant by the moon's 'dark side'. So does the moon really refuse to spin for us? No, the reason for an omnipresent countenance and an eternal 'dark side' is simply a beautiful twist on the nature of twists. It turns out that the moon spins almost exactly one time on its axis for every one time it orbits our planet. That is, a month on Earth takes essentially a 'day' on the moon. The moon and Earth have, over the ages, fallen into this unusual but stable harmony.

It would not be fair to call the match a 'coincidence', but our Earth-Moon relationship could have played out in an infinitude of other ways. Neither *Phobos* nor *Deimos*, the two moons of Mars, conceal a 'dark side' from Martians. Of Jupiter's sixty- three satellites, not one is known for a forever face.

CERES

NOVEMBER 8

This week might be a good time to supplement your star-gazing with even a cheap pair of binoculars. Point them to the south-east at approximately 10:00 pm—particularly on Friday, November 9. Is it a bird? Is it a plane? Neither star nor planet; neither sun nor moon: If you're aiming at the right spot, then that bright singleton in your field is a uniquely large and well-positioned specimen of a new astronomical category: A *'dwarf planet'*.

Ceres, the non-star star of this week's sky show, was formerly known simply as an asteroid. It is the largest of any found in that 'Asteroid Belt' of under-sized sun-orbiters between Mars and Jupiter. Its 600 mile diameter made Ceres quite a large rock, but not large enough to be called a planet—until the Pluto controversy of 2006 forced a facelift on the dictionary. On August 24, 2006, Ceres, Pluto and *Eris* (a distant solar system object that has never been officially classified as 'planet' but has always been larger than Pluto) all donned their new *dwarf planet* identities.

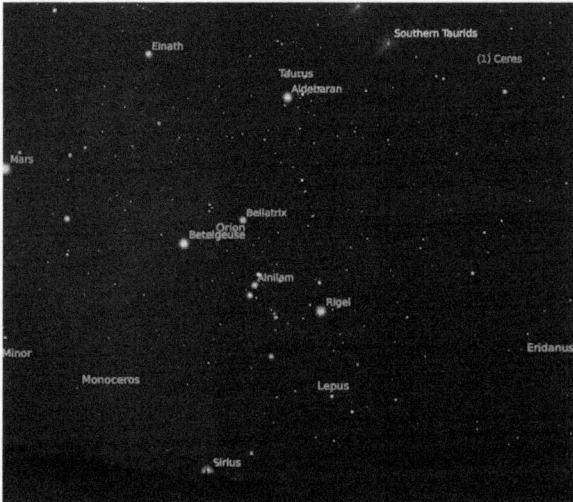

Ultimately and technically, a rock doesn't obtain dwarf planet status by falling short of a particular size. A dwarf planet resembles a planet in many key respects (orbits the sun, does not orbit something else that orbits the sun, maintains a sphere-like shape) except one: A dwarf planet has not 'cleared the neighborhood' of its orbit; unlike a proper planet, a dwarf planet shares its orbital path with other objects and does not dominate.

This Friday will be the first time since its re-birth that Ceres will reach the side of Earth directly opposite that of the sun. This orientation, combined with the absence of moonlight during night viewing hours, makes for a good time to try and spot the rogue.

Where specifically should you look? Use Orion, the colossal hunter low in the southeastern sky, as a reference. Follow Orion's belt straight up about five belt lengths. You will not miss the prominent and orange *Aldeberan* — the 'bull's eye' of *Taurus*.

Now move your focus right (west) the same distance you went up from Orion's belt. You will find yourself in the tail of

Cetus the whale. Cetus is not composed of particularly bright stars. To the naked eye, nothing striking jumps out. In binoculars, therefore, Ceres will be among the brightest points in view. It may be a dwarf, but it lives and shines nearby. Light that comes to us from Ceres to us takes about fifteen minutes to make the journey. By the time Pluto's light reaches us, in contrast, it's been traveling for almost four and a half hours.

AQUARIUS & THE CLOCK

NOVEMBER 15

Two Sundays ago, we turned our clocks back an hour. Although the extra sleep was undoubtedly appreciated, we now feel the deeper impact in our daily lives: This week, the sun rises at approximately a quarter to seven in the morning, but sets at approximately half past *four* in the evening! Our days are now ending dramatically earlier than they had been. Of course, this means that stargazing can begin that much more promptly.

At some level, the *Daylight Saving Time* device can become a habit which seems reasonably responsive to a natural phenomenon: Even without clocks or calendars, the sun automatically shines on us for longer periods during summer days and for shorter periods during winter days. Upon closer scrutiny, however, confusion may set in. We don't appear to be compensating in the expected direction.

Those of us who have adopted *D.S.T.* (and it isn't all Earth-dwellers—much of the non-continental United States and even the non-Navajo portions of Arizona, for example, follow one clock all year) set our clocks *ahead* during a season that is already characterized by later sunsets.

Are we fixing something by breaking it further?

No. Controversial as it may be, the device is not self-contradictory. The theory underlying Daylight Saving Time, adopted during the infant industrial age of World War I, is the following: The bonus daylight we receive when we are tilted toward the sun is more economically useful at the end of the day than at the beginning. During the months that bring both earlier sunrises and later sunsets, many technologically developed nations have decided to take the morning gift and trade it in for an evening double-gift.

By turning the clock forward by an hour, we express our willingness to postpone sunrise from 4:30 am to 5:30 am in order to keep factories (etc.) functioning all the way to 8:30 pm at a lower energy cost. When we recently turned our clocks back, we did not do anything artificial; we returned to "standard" time.

So take advantage of the earlier night-sky! This Saturday, even as early as 8:00 pm, look up to the south-west. Check out the setting 1st quarter moon. Look directly above it approximately three and a half moon-widths.

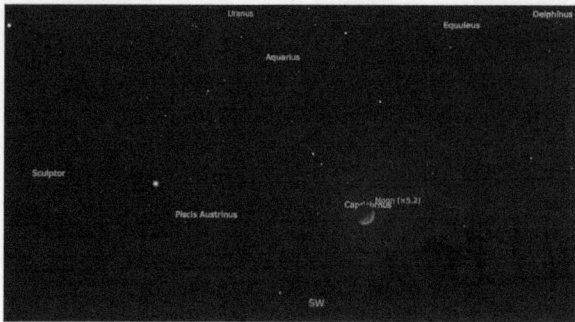

The yellow super-giant you see is *Sadal Melik*, the primary star of the constellation *Aquarius*. It produces light approximately six thousand times brighter than that of our sun. (Our sun is simply a great deal closer.) When you look at Sadal Melik, you stare almost dead-on at the *celestial equator*: the outer-space extension of Earth's midriff. Given that light takes time to get from one far place to another, your current impression of Sadal Melik is actually a snapshot taken approximately seven hundred and sixty years ago. Now *that's* setting a clock back.

ARES

NOVEMBER 22

We are not quite ready for the magnificent 'winter constellations'. Jupiter is getting more and more difficult to catch in the slim window after sunset. Mercury, Venus and Saturn all appear before dawn, rather than after dusk. So perhaps the eternal sky show has reached an intermission. But wait: What is that bright orange newcomer rising in the east at approximately 7:30 pm? None other than our next-door neighbor *Mars*.

This season, Mars hangs out in the constellation *Gemini*—playing third wheel to the twins *Castor* and *Pollux*. It climbs higher in the southern sky each moment of any given night. It also starts its climb roughly five minutes earlier each subsequent evening. So Mars is getting easier and easier to spot as November gives way to December.

December 18, in fact, Mars will reach peak proximity to Earth.

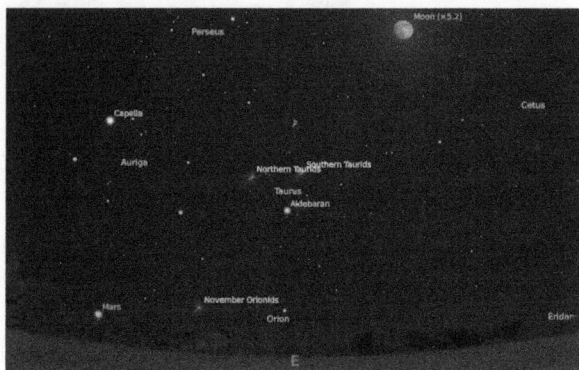

You do not need binoculars to spot Mars. Just wait at least three hours after you catch a sunset and turn half a counterclockwise circle. Your search for Mars can be made even easier if you first focus a bit south of east toward the three stars of Orion's singularly recognizable belt. Orion the hunter is big and low. He occupies an increasingly central place in the southern sky as our New England nights grow colder and longer.

Find the belt and then direct your gaze left (east) and slightly up (north) toward Orion's shoulder. Let your eye rest for a moment on the red supergiant star called *Betelgeuse*. Maybe even pause to consider that the Betelgeuse you see is in fact the Betelgeuse as it existed during Galileo's childhood—or that the Betelgeuse the young Galileo knew was actually the Betelgeuse of approximately 1150 A.D.; even light requires more than four hundred years to traverse more than a couple quadrillion miles.

This month, the ascent of Betelgeuse heralds the appearance of Mars. As super, giant and hued as Betelgeuse may be, it invites your eye to keep scanning left (east) at approximately the same altitude. You'll find only one object with more apparent girth, luminosity and tint than Betelgeuse:

Mars. Bright as it is, the planet Mars is not even remotely the size of a star—let alone a supergiant. On an outer-space scale, it is simply super-close: approximately fifty million miles or four *light-minutes* from Earth.

ARIES

NOVEMBER 29

The sun rises in the east. It sets in the west. From season to season, the arc it carves through the sky smoothly varies in height and slope, but it always peaks near due south—at least to those of us who live above the equator. The sun's east-to-west path across the southern sky is known as the *ecliptic*.

The chief naked-eye objects of our solar system—Sun, Mercury, Venus, Earth, Moon, Mars, Jupiter and Saturn—all turn out to lie approximately in one plane. What does the opening bit of terminology have to do with the above bit of geometry? Just this: When we stand on Earth and look at the sun, it is as though we gaze at a large sky-slice or rigid sheet—edge-on. Our moon and all the planets live essentially on this sheet. But, again, so do we. We do, therefore, not experience a bird's-eye view of our own world. A table-top at eye-level reveals only its boundary: a line. The bounding edge of our egg- shaped cosmo-logical table-top is that curved line called, you got it, the ecliptic.

The ecliptic therefore serves as center-stage to the celestial theater. Note the sun's journey on a given day, and you immedi-ately know where to look for the moon and planets come night. Mars, for example, is expected to rise roughly two hours after sunset this Saturday. This then tells us as much about place as it does about time. Mars is expected to reach the top of the ecliptic by about midnight. At that time, you'll find Mars pretty much where you had found the sun at noon: due south.

Constellations that lie along the ecliptic also make the relentless march from east through south to west—night after night, month after month. These constellations are such reliable night-sky landmarks that they are grouped under their own special category: the constellations of the *zodiac*.

This month, it is *Aries*'s turn. To be sure, if you're thinking about your horoscope, it might well seem to be *Sagittarius* time. This simply means that the *sun* is found in Sagittarius. Sagittarius, consequently, is not visible at night.

Look due south and comfortably up at approximately 9 pm any night this week. In other words, check out an accessible portion of the sky at an accessible hour of the night. Find the two brightest stars toward the center of your southern gaze. The pair will form a short line from the brighter star at top-left to the dimmer one at bottom-right.

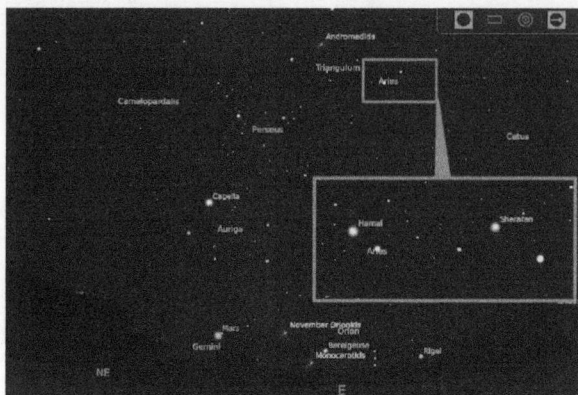

Meet *Hamal* and *Sharatan* of the constellation Aries. They are, respectively, the head and second horn of the mythological ram. In ancient times, the arrival of the sun at the head of the ram signified the arrival of spring. (The sky has shifted a bit since then, so spring now comes when the sun reaches *Pisces*.) In modern times, we understand Hamal to be a sun in its own right— approximately 4 times more massive, 18 times larger and 55 times brighter than our own.

READY FOR RED

❧

DECEMBER 13

*W*inter weather may be characterized by white; Christmas décor may emphasize green. This December, however, if you're not wearing red, then you're clashing with the universe. Indeed: It's *Mars* time.

A couple of weeks ago, we discussed the gradual introduction of Mars from the eastern horizon. We are no longer in exposition. In more ways than one, Earth's ginger- hued neighbor is about to vault into the center of our viewing field.

Tuesday, December 18, first of all, Mars

will reach its closest approach to Earth. In outer space, closeness is relative: Our approximate separation of a mere 54.8 million miles—a four minute errand for both the there-to-here red and here-to-there blue beams of light—makes siblings out of friends. The Mars-Earth gap can grow to almost five times greater than it will be this December, so proximity differences really do make a difference.

Christmas eve, moreover, Mars's presence will bear two presents. While it will continue, for all terrestrial intents and purposes, to be characterized by closest approach, Mars will also reach 'opposition'. This means that the Sun and Mars will arrive at precisely opposite sides of Earth. Like our moon when it appears full, the greatest possible portion of the Martian surface will reflect light straight toward Earth—and at the darkest possible Earthly hour. Put even more practically, Mars will be visible and distinct essentially all night.

This month, look for Mars in the zodiac

constellation of *Gemini*. That is, in the vicinity of 10 pm, look up and toward the vicinity of due south. On December 18, for example, find *Betelgeuse*, the left shoulder of Orion the hunter, and turn your head up and to the left. These instructions can be offered and regarded carelessly because Mars will be so uniquely bright.

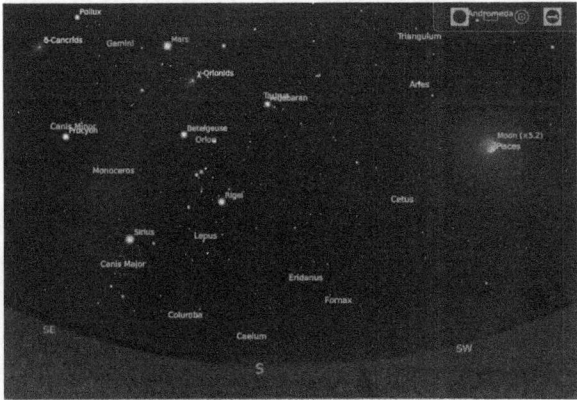

Even on December 23, when Mars will compete for attention with the full moon — down and to the left about three moon-widths from the planet — it will still shine stronger than *every* star in the sky. . . including Sirius, Orion's faithful companion. The dog-star is the brightest star in the northern sky and usually even more intense than Mars. But not this time. Ho ho ho.

SOLSTICE

DECEMBER 20

This Saturday marks the astronomical beginning of a new season: December 22 is the winter solstice. Although a solstice is not an especially visual event, its semi-annual arrival celebrates one of the relationships at the heart of astronomy's appeal.

The deep sky is where time and space comingle: the position of the sun is a tick on our clock, the shape of the moon is a page on our calendar. We embrace this delightful confusion at the start of both summer and winter: A solstice is, by definition, both a particular point in time and a particular point in space.

At 1:08 am (Eastern Standard Time) on Saturday, the rays of our sun will slam directly and perpendicularly on the *Tropic of Capricorn*: a line of latitude essentially 23½ degrees south of Earth's equator. This is as low on the Earth as the sun's rays ever point.

If you were to visit the tropic and lie face-up on the beach at noon this Saturday, you would find the sun all the way at the tippy-top ('zenith') of the sky. Its beams would shine straight down on your forehead. Winter trips to 'the tropics' (that 47 degree band between Cancer and Capricorn) do, after all, make sense.

For all other latitudes on Earth, the sun will be lower in the sky and its rays will graze rather than attack Earth's surface. For us here at approximately 42 degrees above the equator, we ought not look straight up nor lie down in order to find the sun—even if the ground happens to be clear of snow.

The sun simply will not climb any higher

than 25 degrees above the horizon (that is, 90 - 42 - 23 degrees). And that's at a mid-day peak. This low altitude and shallow angle becomes a nadir for the entire year.

Since the sun has to reach this (minimal) maximum via a gently curving journey across the southern sky, it cannot embark from anywhere terribly far east nor end up anywhere terribly far west. To stay smooth, in other words, a low and bounded arc must be a short arc.

So the day of the winter solstice is one that bring us fewer minutes of daylight than almost all other days of the calendar year. Due to the artificially un-smooth character of time zones, among other details, the winter solstice does not **perfectly** coincide with latest sunrise nor with earliest sunset.

On Saturday, the first day of astronomical winter, the sun will rise at 7:19 am and set at 4:24 pm. Sunset time will have already reached an earliness peak back on December 15, when it set at 4:21 pm. Sunrise will not reach a lateness peak until

January 9, when it will rise at 7:22 am. Each of these events will occur dramatically south of due east and due west, respectively. Picturing the sun's position in our sky, rather than counting minutes of daylight, thereby provides a fundamental way to grasp the meaning of a solstice.

For sunlight to hit us in the above fashion, the sun itself must aim from a specific location. This position, currently found way in the western portion of the constellation Sagittarius, is an established and defined point. That point is literally called '*the*

winter solstice.' When the sun is way up there, the trajectory of its rays inspire candles, coats and cozy-nesting way down here. Unless, of course, 'way down here" means way 'down under'.

COMET 8P/TUTTLE

DECEMBER 27

Christmas eve, Mars was kind enough to play Rudolf in the low southern sky.

Might he have dragged other familiar figures over from the north pole? Well, yes: As long as we define both 'familiar' and 'north pole' a bit liberally.

If you're looking for something peaceful to do before midnight on New Year's Eve (or on any clear night in the week surrounding), walk out and look up. *Comet 8P/Tuttle* will have just flown in from near the north celestial pole. And boy is his tail tired.

A comet is, in one key respect, like a planet: It orbits the sun. Comet 8P/Tuttle, in particular, takes almost fourteen years to make one trip around the sun. For much of its journey, it is not visible from Earth. The comet's path does not lie in or even near the *ecliptic* plane that serves as approximate home to the all planetary orbits.

Comets such as 8P/Tuttle are often referred to as 'dirty snowballs'. They are composed of rock, dust and 'ice'. At Fahrenheit temperatures of less than 450 degrees below 0, however, the 'ice' recipe involves more than frozen water; Gasses such as carbon monoxide, carbon dioxide and ammonia solidify and pack together to help form a nucleus of, in 8P/Tuttle's case, roughly nine miles in diameter. The surrounding material, or 'coma', will then span a far greater distance.

Comets such as 8P/Tuttle are identified by a 'tail' or actually pair thereof. The primary tail does not trail behind a comet in the manner of superman's cape nor, for that matter, the pom-pom on Santa's cap. There is no air in outer space, so wispy accessories are subject to the drag of no conventional wind. Electrically charged particles in a comet's coma, rather, are pulled by the sun's severe electromagnetic influence—the *solar wind*. So a comet's ion tail points toward the sun instead of flopping away from the direction of travel.

This New Year's week, January 2nd in particular, Comet 8P/Tuttle will make a close approach to Earth: approximately 23 million miles. This is just close enough, if the sky is extremely clear, to make Comet 8P/Tuttle visible to the naked eye.

Look high up in the southern sky: so high that you're almost looking north. The Comet will be at the eastern edge of the constellation Pisces. If you identified Mars on Christmas eve, then just look above Orion and find it again. Turn your gaze to the right until you pass the eye of Taurus the Bull and pause at the dazzling 'Seven Sisters' or *Pleiades*. If you continue to follow this west-headed line essentially the same distance you just traversed, you'll land at Comet 8P/Tuttle. Make your welcome warm; that is one cold voyager.

MERCURY RETURNS

*T*hey say that we should burn the message, not the messenger. Sage advice when the news is bad. This week, however, the message is exciting and the messenger is—at least during the day—already plenty hot on his own. Indeed: The planet *Mercury* has just returned to the night sky. For months, it has been visible only during the dawn. Now, having completed its most recent crossing from the right to the left side of the sun, Mercury can be observed in the low south-west at twilight (these days, just before 5 pm).

Thursday, January 10, for example, look up just as it gets dark. You will find an elegant crescent moon sinking in the south-west; it heads toward the same horizon that had swallowed the sun minutes earlier. Between the moon and the ground, along that smoothly curving sun-path known as the *ecliptic*, you will notice one — and only one — bright point. This is Mercury.

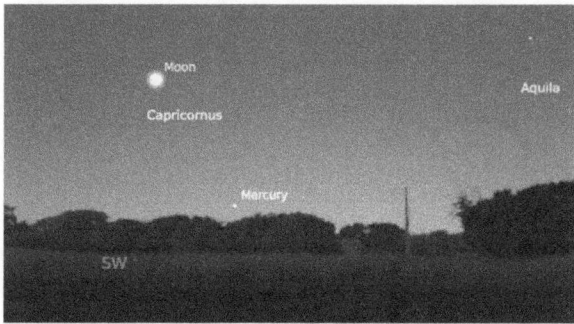

Each night thereafter, the diminutive rocky planet will be higher in the sky for a longer time period after sunset. As the moon waxes in phase, however, it climbs up and eastward in the sky at a far greater rate than Mercury. The two will therefore drift apart from night to night. This separation, on the one hand, will lessen the lunar wash-out of Mercury's glow. A nearby moon, on the other hand, can serve as a

helpful landmark in the endeavor to identify planets.

As he pushes east — away from the setting
sun — Mercury will spend a full week
increasing in prominence. On January 15,
it will remain in the sky an hour and
twenty minutes after sunset. It will
continue to trek up and out, but will cease
to get any brighter. Why? Because, alas,
Mercury's steady mellow sheen is
distributed, not produced, by the rock.
Much like our moon, Mercury is simply
not a star; it is *not* a perpetual explosion of
gas and radiation. It has a solid surface that
reflects sunlight.

As a consequence, again like our moon, the
portion of Mercury's surface that appears
illuminated to us depends on the angle
formed among Sun, Mercury and Earth.
Mercury, that is, undergoes phases. From
January 15 through January 21, Mercury
will wane from gibbous (three-quarters) to
just about a half-disc. The shrinking shape
itself is not visible without a telescope, but
it does result in a slight fading.

So try to catch a glimpse of the 'messenger planet' at some dusk over the next two weeks. If you have binoculars, check out that creeping shadow. But simply delight in the shadow's message; don't seek too much information. Mercury is a celestial body, not a groundhog, for heaven's sake.

⁂

JANUARY 24

*O*ne of the most reliable and rewarding gifts of a clear January sky is the 'Winter Ellipse'. The name does not refer to one proper constellation. The Ellipse comprises, rather, the prominent intersection of six officially recognized star-groups. The uniquely bright and conveniently positioned envoys from each of these collections end up forming their own renegade shape or *asterism*. The shape, an imperfect oval, merits discussion because it is just so darn large and in charge throughout frosty dark months.

On any night you please, face due south.

Look up—but not so high as to strain your neck. The earlier the hour, the more you should twist your head to the left (east). For example, if you happen to be standing outside at approximately 9 pm, do not bother to swivel at all. If it's more like 8 pm, start looking toward left-center rather than toward center field.

With no prior star-gazing experience at all, you will soon see *Orion* the Hunter. The linear star-trio of his Belt is the giveaway. Had Orion happened to have donned suspenders, I would have precious little idea how to orient myself in the southern sky. Orion's right knee (which may easily strike you as a foot) is *Rigel*. Generally the brightest star in Orion and the sixth brightest star in our entire sky, Rigel is a 'supergiant'—approximately seventeen times as massive and forty thousand times as luminous as our sun. Rigel appears so bright because, unlike some other stars, it truly is.

Its light rays enjoy ample opportunity to dim spread during the **eight hundred**-odd years they spend traversing the distance between Rigel and Earth. Yet they're anything but dim. Follow the ellipse clockwise (down and right) from Rigel until your eye hits the next singularly vivid point. Meet **the** brightest star in the northern hemisphere's sky. No joke: It's *Sirius*. In direct contrast to Rigel, Sirius makes its incomparable impression by living so nearby.

Light zips from Sirius to Earth in under nine years. So Sirius out- dazzles all competition despite boasting only twice the mass of our sun. Sirius is the primary star of the *Canis Major* constellation. Orion, in other words, hunts softly but carries a big dog.

Let your eyes continue clockwise to Sirius's companion, *Procyon*. Procyon shines for *Canis Minor*, the Little Dog. Above and to the left of Procyon are, in clockwise order, the twins of Gemini: *Pollux* and *Castor*. These stick-figure brothers (usually visualized as leaning leftward) consist of a number of variously massive stars, but at least three of the stars are known to govern their own planetary systems.

At the top right of the ellipse rides *Capella*, front-running for *Auriga* the Charioteer. Finally, we come full-ellipse with *Aldebaran*, the eye of Taurus the bull. You'll know you've correctly traced the ellipse if you find the bull's-eye along a continuation of the line defined by Orion's Belt.

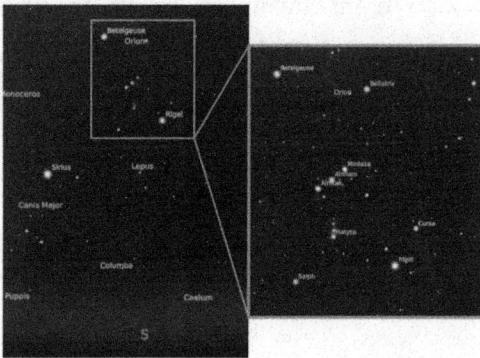

If an ellipse is an egg-shape, then the yolk (interior) is Betelgeuse: that red supergiant who constitutes Orion's left shoulder. Once you look, you can hardly miss Betelgeuse's carroty hue. What your naked eye cannot tell you, however, is what the term 'supergiant' means in the context of Betelgeuse's volume: If Betelgeuse were located in our sun's current spot, then the orbits of Mercury, Venus, Earth and even Mars would all be swallowed up in its girth. Orion, in the final analysis, is in stellar shape.

He's got a knee and mid-section that are quite easy on the eye. He dares you, more-over, to knock that fire off his shoulder.

JANUARY 31

*L*ast week in this column, we met the *winter ellipse* — a great southern sky *asterism* (unofficial star group) that will anchor us during the cold, dark nights of February. But a flurry of shorter-term activity once again reminds us that mother nature performs magnificent matinees; pre-dawn shows, such as this week's, are not to be missed.

First off, the moon now 'wanes' (appears to shrink). It hit its 'last quarter' phase (which means it looks like the left half of a circle) Wednesday, January 30; it will be new (entirely dark) again next Wednesday, February 6. Whenever the moon experiences the waning half of its monthly cycle, you can expect to find it in the morning sky rather than at night.

Second, some of the more dazzling planets now gather in the mornings. Venus expresses its identity as the 'morning star'. In ancient Athens, this 'bringer of light' was known as *Phosphorus*, while the brightest 'evening star' was *Hesperus*.

Eventually, Greek astronomers accepted the Babylonian view that the two were in fact one: one object that wandered back and forth across the otherwise fixed canopy of stars. The Greek word for 'wanderer' is *planet*. Hence, the club was formed: Venus, Jupiter, Saturn, Mercury, Mars, Sun and Moon were the seven extraordinary creatures that had somehow cut their tethers to the thousands of dots comprising immutable constellations. Since the development of telescopes, the rules for membership in this club have gotten a bit more complicated. Just ask Pluto how he feels about the details.

So, you can expect to see Venus rising elegantly in the southeast around shortly after 5 am this week. Right about the same time, particularly on February 1, Jupiter does the same. If you had to pick one day to check out these lustrous wanderers, beat the sun to breakfast on Friday. Venus, Jupiter and the moon will all float together —in a super- skinny triangle extending from left to right (with Jupiter forming the dimmest vertex). If you notice a fourth (less intense but still quite prominent) object just below the moon, then you have met *Antares*—the brightest star in the constellation *Scorpio*.

Antares is about fifteen times the mass of our sun and roughly 700 times the diameter. Its role as 16th brightest star in the northern sky derives from this impressive size, not from proximity. Light takes approximately 600 years to travel here from Antares. And light, were it to travel in circles, would whip around Earth's circumference close to eight times in one second. So Antares isn't exactly our neighbor. He's just a very good friend.

TOTAL LUNAR ECLIPSE,
PART II

❧

FEBRUARY 14

This Wednesday, February 20, New Englanders will be treated to a total lunar eclipse—the last one until late December, 2010.

Last week's column endeavored to explain some of the quirks and quiddities underlying lunar eclipses in general. Here, we provide a quick user's-guide to observing the particular instance on deck for this season. The most practical way to proceed, oddly, is at the middle: The totality peak will be at essentially 10:30 pm. Here, the moon will be entirely engulfed by Earth's shadow. A dark disc will sit in the southern sky at higher than 60 degrees from the eastern horizon (where 90 degrees refers to due south). The black disc, previously a ruddy-hued moon, will be flanked by *Saturn* to the lower left and *Regulus* (brightest star of the *Leo* constellation) to the upper right.

This time is not precisely exact, but it is exactly true. That is, this is not a weather report; it is more than a best estimate regarding probabilities and extrapolations. True, the true peak is one specific moment that will neither register on your watch as the dot of half-past nor will occur to your eyes as sharply distinct from any of the many moments surrounding it. Yet, it simply *will* happen—and as close to the expected time as you would ever wish from any commuter train.

Unless, of course, the weather does not cooperate and clouds get in the way. Even then, the show will go on. It's just that your ticket won't get stamped.

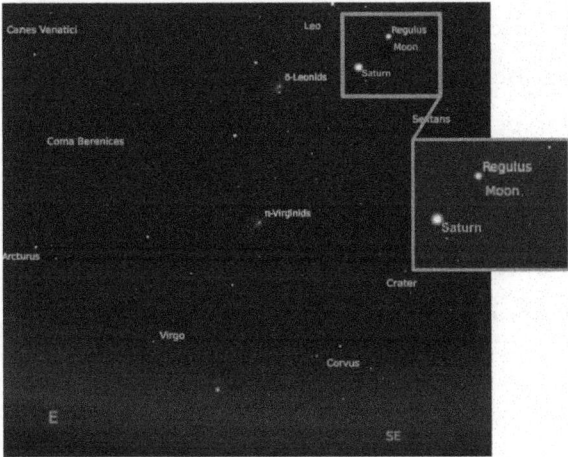

Earth's shadow will begin creeping over the full moon as early as 8:43 pm. The full moon will already have risen in the southeast over three hours prior. If you face the south and look left anywhere around 9:00 pm, you will see what it could be like to squeeze two weeks of celestial time into less than two hours. That is, the moon will wane from full to gibbous to half to crescent to void right before your eyes.

By 10:01, totality will have begun. The moon will sit buried in black until 10:52 pm. At this time, the moon will begin to claw its way out of shade. During this accelerated waxing stage — lasting until ten minutes past midnight—the moon will also be inching its way toward Saturn. From our perspective, the moon will be trekking west—rightward. Even though Saturn is on the moon's left, it, like everything in the southern canopy, spends its nights moving west. It just does so 'faster' than the moon. So, as the moon emerges from obscurity, it will simultaneously bridge the gap to the ringed gas-giant.

So Wednesday night, give yourself three belated Valentine (or President's Day) gifts: a thermos of cocoa, a wool hat and a total lunar eclipse. If you over-think it, you might decide that an eclipse is just a simple film of two vying circles and some light-dark contrasts. But opposites attract. And Hitchcock movies drew their aesthetic appeal from something more than the stars.

PART II

INTERLUDE: A BRIEF BEDTIME
ALLEGORY

AND STILL IT MOVES

A TRIALOGUE

APROPOS OF WHAT?

The following is a brief narrative originally intended to stand in isolation as just that: a short story. The story is, however, short even for short. It blatantly lacks, moreover, in some truly basic story components. Nothing in the account ever really amounts to a conflict, for example. Nothing in the account, in fact, ever really amounts to anything nor happens, really, at all. Some people look at some stuff; we hear about it. Some kind of defense (or at least plea) seems warranted (or at least invited). For this, we offer a choice of two. The offer is not, in the symbolic logic sense of the term, exclusive: You are welcome to choose both.

First: Call it what you will, the remaining set of pages is offered as a gesture of affection. Its primary purpose is to leave the reader with a moment of uncomplicated contentment, ideally one which skates through and manages just barely to register in consciousness as net marginal uplift We hope to share a bit of pleasantness with enough pluck to exceed or survive that which these days seems to be the increasingly overwhelming or under-performng costs of reading.

For the near-violent exacting of such costs, contemporary documents that are not non-fiction tend very much to be not without blame — particularly those feral texts which want for links to other web-pages, fields for userform data, options for express delivery or Likert Scale rankings for average customer satisfaction. Here, an offer of negligible profundity is extended, but, we actively assert, at a correspondingly low low price to the soul.

Against facilitating, prompting or partially contributing toward, through words, the experience of something resembling pure diversion, the odds seem astronomically high.

Hence, *second*: This narrative comes with a fallback option — a truck-driving day-job to support its predictable failure to fly.

In the event of academic emergency, the pages below stand ready to serve as the manual for an introductory yet highly particular semester-long laboratory course in scientific star-gazing — some assembly required. To the extent that necessity is the mother of invention, astronomy might well be considered the womb from which emerged the study now called classical physics. To the extent that a child is father to the man, it might be noted that some key structural elements of the vignette to follow are pilfered straight from Galileo's 17th century satire *Dialogue Concerning The Two Chief World Systems*. If by some miracle, this thing passes for any parent as a legitimate bedtime story, however, then all the foregoing should be loudly ignored.

To the best of the author's knowledge, every illustration or image included herein is elsewhere identified as belonging to the public domain.

Excepting a depiction of the *Sun* (presented as an anthropomorphic caricature) and that of *Venus* (presented as an 'evening star'), all the images to follow are reproductions of sketches made by the great Galileo. They were made during or shortly prior to the year 1610 A.D., for use in his self-published work, *Sidereus Nuncius*.

Each sketch is of what Galileo saw. In

order to see, Galileo looked over and over again, clear night after clear night after night. If the nights were not clear, then he waited and waited and waited.

He saw how the sky looked when magnified to unprecedented levels by a generally unfamiliar innovation called a telescope. Even before Galileo arrived on the scene, a great many eyes had been corrected for nakedness: There were scholars and aristocrats and others who wore glasses.

For a good while before the contributions of Galileo, it had been possible to make and use a lens in order bring the focal point of a human eye closer to an object of interest. There were people who were not unaccustomed to the experience of seeing something as though it were nearer than it actually was.

But then there came a surprising insight requiring an intricate technique: Rather than aiming directly for an eye, the refractive capacity of a lens could perhaps be trained on yet another lens. If curvatures and distances were held to sufficiently strict correspondence, optical effects might well experience, it seemed, amplification. This difference turned out to make a big difference.

Galileo did not invent the double-lens device, i.e.: the telescope. As an early adopter, however, he was a virtuoso. With his own hands, he ground lenses and fashioned one of the first and most powerful versions of the tool. He put it to use in a highly original context. It did its job with unprecedented precision.

Galileo saw celestial bodies as if they had suddenly carried 4 to 10 times closer to his eyes.

～

Galileo believed himself truly
to be seeing things

which no people at no point
in all of human history had ever before
believed themselves to be seeing.

That's a very strange thing to believe
under any circumstances —
but most especially, perhaps,
when it's true.

SUN

*J*esse, Missy Jean and Professor Piyopiyo lived on a land called *'Earth.'* Earth was a big land, divided into many smaller places. Their place was called ' *40 degrees North of the Equator; 74 degrees West of the Prime Meridian.'* Their time was called *'The first day of Autumn.'*

*O*nce again, it was late at night. Thrillingly late. So late somehow that a little later and soon it would be early. In the morning, again, way before the afternoon. It was never fully clear how that part ever fully worked. Once again, Jesse, Missy Jean and Prof. Piyopiyo were in the laboratory on 4, near the corner of 10 and 59, in the building named T of John J. Once again, Missy Jean was at the counter working on a secret potion. This time, thus far, the potion involved sodium bicarbonate, yoghurt, ice chips and a few iron filings. The potion did not yet do what it was supposed to do. It was, however, unusually magnetic for something so tasty.

*J*esse was on the floor studying how rattles and small stuffed animals behaved when he did not put them in his mouth. Jesse had very recently started to make observations under this challenging condition. His observations were still limited. He was fairly convinced that stuffed animals, when kept away from his mouth, were neither tasty nor magnetic.

. . .

*P*rof. Piyopiyo was at his desk. Without looking, Missy Jean just knew it. And she knew what he was doing: scribbling symbol after symbol—all somehow neat and yet unreadable at the same time—in that thick brown journal of unlined paper. He called the symbols '*math*', but Missy Jean knew better.

*M*issy Jean had learned *math* last year in her school. Her school was called 92-Street-Y. Her class had finished all of math in time for summer and she even got her big-sticker sheets right. So Missy Jean knew what math was all about: numbers. Whatever symbols covered Prof. Piyopiyo's unlined sheets, they certainly were not numbers.

*P*rof. Piyopiyo looked at his watch. He looked at Jesse and he looked at Missy Jean. He said,

" Let's go outside and see something beautiful."

*N*obody wished to argue this plan. Missy Jean put on her shoes. Prof. Piyopiyo scooped Jesse off the floor. They went outside, hailed a cab and stood on the edge of the Drive called F-D-R.

*T*he three looked off to what Prof. Piyopiyo called the '*East*'. A sphere of blazing crimson was rising magnificently behind three buildings near two bridges that came from one short place. The place melted off into a mysterious island that Prof. Piyopiyo called '*Long*'. Right now, it was too long to see and too wide to cross, but it all grew out and left, if Jesse heard Prof. Piyopiyo right, from a blue part called the '*Sound*'.

· · ·

*J*esse was a little confused by this name; all the ups and overs and outs and colors seemed so nice—so slow and silent. Cool and quiet, not like a sound, but also not like a picture. Not at all like one of those pictures you could hold in your hand on a buzzing phone. No, watching the sunrise was more like sitting on the couch with Missy Jean and a peaceful movie—but lost in cushions made of clouds and no reasons to rewind. Things were calm and clear with Missy Jean. She even knew how to spell movie—'D-V-D'—and how to take care of him. He was proud of her for that.

*A*t that moment, at least, things were not unclear to Missy Jean. But neither were they clear. They just were. She had stopped listening to Prof. Piyopiyo or to anything else from the moment she started looking. Her mouth hung slightly open. She did not hear Prof. Piyopiyo when he told Jesse that the fiery circle was called the *'sun'*. She did not, at that moment, need to know whether the treat was called *'sunrise'* or *'runsize'* or *'google-plex'*. It was a treat no matter what anybody wanted to call it.

*T*he three remained quiet for a long time after the stars faded from view and the sun dominated the low sky. The whole thing was indeed a movie. Both the fading stars and the rising sun trekked from one sky edge to on their way to the other. It appeared as though Missy Jean, Prof. Piyopiyo and Jesse were under a big round tent. Somebody invisible was pulling the tent top from left to right. At last, it dawned on Missy Jean:

" It might be nice to see that again some time."

*S*he paused.

" But some other time, I think. I'm too happy right now."

Jesse was also happy. He was happy for the sun. He was sure the sun would get some rest and burble up again some other time. Jesse, Missy Jean and Prof. Piyopiyo went home and ate and played and napped and played and ate.

MOON

*B*efore pajama time, Jesse was aware of the dusty, dusky traces of sunlight. Were they coming or going? At the lab, Jesse found Missy Jean. She too was looking forward to another sunrise. Finally, Prof. Piyopiyo said,

" Let's bring our suppers outside and see if the sun rises again."

*A*nd so they did, but so it didn't.

. . .

*I*nstead, they saw the moon— far higher than the place in which they had recently found the sun. The moon was bright and beautiful but looked like someone had bitten it. Half it was missing.

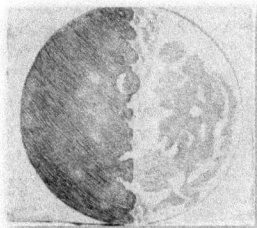

*W*hy?" thought Jesse.

*H*e remembered one night when the moon looked like a perfect circle. He remembered it well because the moon kept following him around that night —even when he was in his car-seat zooming up that same FDR Drive. What kind of creature would eat his moon?

*J*esse thought and Missy Jean asked. Prof. Piyopiyo assured them:

" Don't you worry. The moon's still there. Part of it is just hidden in shadow. The moon cannot make its own light the way the sun can.

" Sunlight bounces off the moon and heads toward our eyes. Sometimes more gets bounced, sometimes less. Depends where the moon is."

*S*upper was long, leisurely and quiet. Missy Jean was deep in her mashed potatoes. Jesse was deep in thought: How do some things *make their own light*? The sun did, but the moon didn't? Did he? Did Missy Jean? Prof. Piyopiyo probably did: He certainly made a lot of his own noise.

STARS

"What about the stars?"

*a*sked Missy Jean.

"Why can't we go say hello to them before we say good night to ourselves?"

"Why not indeed?" replied Prof. Piyopiyo.

*T*he three walked outside and Prof. Piyopiyo pointed up.

" See that sideways V that looks like a mouth? It's head of a bull named '*Taurus*'. That bright red giant star, Aldeberan, is its eye."

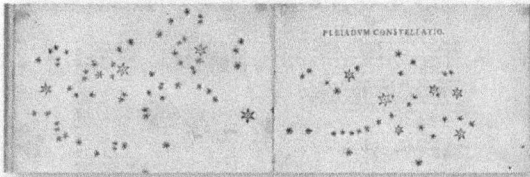

*M*issy Jean and Jesse loved animals. They kept their eyes on the bull for a long, long time. Long enough, it turned out, to notice that it, like all the animals and fish and twins and hunters and everything else around it, it seemed, slid smoothly across the sky— from down and left to up and center to down and right, it seemed.

Good job, world," said Missy Jean,

Just keep on turning around us like that."

Yes. Good job, world," thought Jesse.

Good night, you two," declared Prof. Piyopiyo,

The early bird catches the worm".

*M*issy Jean was not so interested in catching a worm, but Jesse had no strong feelings on the matter. His eyelids dropped fully by the word *'bird'*. Somehow, he later found himself in bed.

VENUS

FRIDAY

The next evening, the three rushed outside and zoomed over to the Hudson River to see if they could catch the sun setting in what Prof. Piyopiyo called the 'West'. There was a big tugboat blocking their view. Sadly, it seemed that they had arrived too late.

Happily, however, they arrived just in time to see the biggest, brightest star they had ever noticed. It seemed be setting—just like they had been hoping for the sun.

" Wow; dazzling," thought Jesse.

" Is that super-duper star always to the right of Taurus?" asked Missy Jean.

" Actually, no," explained Prof. Piyopiyo,

" That thing wanders back and forth--but always stays near the sun. It wanders past the bull, the ram, the scorpion and the other nine animals who march together across the southern sky. Because it wanders, it's actually called a *'planet'*."

It's the planet Venus," finished Prof. Piyopiyo.

*J*esse suddenly felt different from how he had a moment ago. He had been excited-happy and now he was calm-happy. It was funny how having a name for something memorable seemed to solve a mystery. It was also funny how a mystery's solution could create its own confusion: Which was more fun — searching or finding?

JUPITER

THURSDAY

*T*he next night involved more searching and more finding. The three went out far later than they had on previous nights.

*A*fter this evening's dinner (juicy chicken to eat and chickeny juice to drink), they played and sang and danced and danced and sang and played for tons of time before heading out to the edge of the valley. The sun was long gone; Venus was long gone. Even Taurus was getting close to that mountain in the west.

.   .   .

*B*ut up in the sky, amongst the connect-the-dot crab that Prof. Piyopiyo called '*Cancer*', blazed another unusually bright star—almost as big and beautiful as Venus.

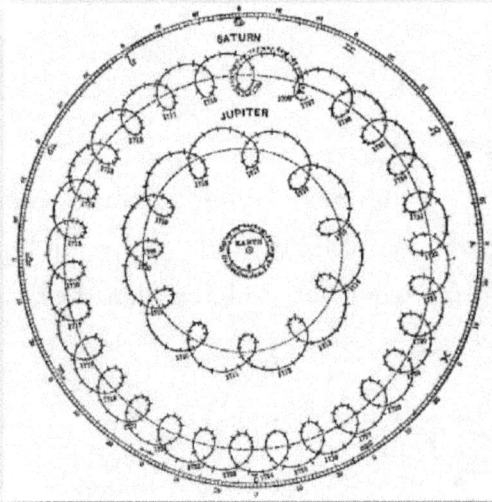

*N*either Missy Jean nor Jesse failed to notice. How could they?

• • •

*P*rof. Piyopiyo did not fail to answer their unasked question:

66 You got it, guys: Another planet. That one is Jupiter.

66 Planets aren't always that bright, but they do always dance to their own tune. Some day, some year, you won't find Jupiter among the crab stars. You'll find it among the lines of the lion or the specks of the sea-goat."

PHASES OF VENUS

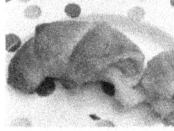

(BINOCULARS RECOMMENDED)

The next and final evening, Jesse, Missy Jean and Prof. Piyopiyo rushed out early in order to try and catch the sun before it dropped behind the big mountain. Again, they missed it. It was too bad, but not too too bad. Neither Jesse nor Missy Jean truly wanted to win in a competition against Mother Nature or Father Time.

. . .

*A*ny disappointment they might have felt disappeared the moment Prof. Piyopiyo pulled two small but clunky black objects from his bag. Each object was a pair of binoculars. Jesse and Missy Jean each gazed toward the mountain through the long, thick glasses. Suddenly, it seemed as though they had stepped a dozen times closer to outer space.

*W*ithout the binoculars, they spotted and recognized Venus. With the binoculars, a bit of confusion set in.

Missy Jean spoke:

" Now Venus looks like the moon!"

*J*esse knew what she meant. Now that he was closer to Venus, it no longer looked like a wandering star. Like the moon, this circle was missing a chunk.

*S*o some sky-wanderers did not make their own light?

*P*rof. Piyopiyo PhProf. Piyopiyo nodded as though he had heard Jesse's thought. Instead of answering, he asked a different question:

"Missy Jean, do you think the firm ground below you makes its own light?"

*M*issy Jean replied with confidence,

"No silly; dirt isn't fire!"

"Makes sense to me," stated Prof. Piyopiyo.

MOONS OF JUPITER

(A SECOND AND FINAL CALL FOR BINOCULARS)

*I*ndeed, at some times, some space things were starting to seem clearer to Jesse. Yet he was surprised to find—at some other times—that the more he looked at some other other things, the less clear the other other other things seemed to seem. If some sky-wanderers did not make their own light, and if Earth did not make its own light, did that mean Earth was like a sky-wanderer? It certainly did not feel that way.

. . .

*I*f circles were shapes, for example, did that mean that shapes were circles? It certainly did not seem that way.

*Y*et something about those wanderers made Jesse feel a bit dizzier in his mind and even dizzier on his feet.

*T*o make matters more surprising, Prof. Piyopiyo then pulled four objects out of his pack: three large and one small, but all squishy. Jesse loved that pack. He loved how it was bigger on the inside than it was on the outside.

*P*rof. Piyopiyo unrolled the three large objects. They were sleeping bags.

. . .

*B*ut neither Missy Jean nor Jesse noticed until later. They noticed the small squishy bag. It had marshmallows.

*P*rof. Piyopiyo explained:

Let's make a fire, roast marshmallows and sleep under the stars. I'll wake you when Jupiter comes out."

*T*hey did and he did.

*J*esse and Missy Jean found Jupiter. Jupiter wasn't quite as sharp as Venus, but Jupiter was bold. Jupiter wore an orange robe. Jupiter was king. It sat solemnly on a floating throne.

hey did and he did. Prof. Piyopiyo offered last night's *binoculars*: those long black eyes that had shown them Venus's missing piece. Jesse twitched. Sure, he liked finding new things that were there—especially when they were hard to see. But last night's hollow bite from Venus was still eating at him. How had Prof. Piyopiyo pointed out something that *wasn't* there — something that before had seemed so easy to see?

They clutched on and took giant steps toward Jupiter — through their new and superpower eyes. When they stared at Jupiter, it was more and so seemed closer. But when they stared at its black surrounding realm, that too was more and yet seemed farther. Their balance grew weaker; their vision grew stronger.

Soon, in a line surrounding Jupiter, four brand new stars fuzzed into view.

Wow," exclaimed Missy Jean, delighting in her discovery,

Oy," thought Jesse.

Are those regular stars or sky-wanderers?" asked Missy Jean, sorting out her discovery.

Prof. Piyopiyo replied:

> Oh, those four wander all right. But they don't just wander around. They wander around Jupiter. Those are four of Jupiter's moons."

A world of worlds going around a world—inside the biggest world of all?

*J*esse's thoughts swirled. They whirled.

*J*esse thought he had a funny feeling, but he did not feel he had a funny thought.

So, a thing in the sky can go around a thing in the sky that goes around a . . . ?" cried Missy Jean.

She slapped a hand on her own mouth—to keep at least her sentence from running away toward forever. So much was slipping. Super-slick yet somehow steadily.

And so, a heavy rock can fly like light—and wander through the dark of space?" wondered Jesse.

The Earth feels so still, and ...," started Missy Jean.

...and still it moves," saw Jesse.

~

And as Jesse thought this thought,
he, too, moved:
this time through time. . .
toward the future.
He grew older.
Perhaps, you might say,
but only by a minute or two
and only at the speed of life.
Yes, but this time, he felt it.

PART III

THE SUMMER BEFORE

EPILOGUE: COUNTER-CLOCKWISE

THE WHAT-NEXT AND THE WARM-DOWN

I. SIMULATE

*D*ownload, install and open any one of several widely available varieties of 'planetarium software package' — whichever appeals to you by whatever metric, i.e.: whichever seems right away to suggest your preferred balance of aesthetic appeal, ease of use and richness of information, i.e.: gotta get gut.

Options range from unapologetically steep retail to pure open-source. For the type of use assumed here, any option you find easily will deliver accurately and suffi-ciently. Reliable options include:

(A) *Stellarium* 0.18.2:
open source, i.e.: free of cost, free of criminality.
https://stellarium.org/en/

(B) *Starry Night Pro 8*, Simulation Curriculum Corporation:
commercial, range of packages: $49.95 - $249.95
https://store.simulationcurriculum.com/

(C) *Redshift 8, 3-Dimensional Space Simulator*:
commercial, starts $49.95
https://www.redshift-live.com/ext/en/shop/products/40517.html

(D) *Celestia: Real — Time 3-Dimensional Visualization*
open source, i.e.: free of cost, free of criminality.
https://celestia.space/

∾

II. ORIENT

Set *location* to *42.1° N, 73.4° W:*

Sheffield, MA, 01257, USA.

*S*et *date/time* to *midnight,*
September 23, 2007.

~

*D*awn:

Begin to prepare.

~

*U*nder *View*, select (turn on, set default to visible):

- *Stars>*

Constellations> Zodiac> Labels

- *Solar System>*

Planets, Moons

~

*U*nder *Guides*, select (turn on, set default to visible):

- *Ecliptic*,
- *Celestial Equator*
- *Solstices*
- *Equinoxes*.

~

'*V*iew from' a location on the surface of '*Planet Earth*',

in particular:

the latitude and longitude specified (above), Sheffield, MA.

~

'*F*ace' *due South*.

. . .

*Z*oom in and out until the horizon is unambiguously (perhaps unremarkably) visible in the frame;

*T*wo large arcs should be clear and distinct in the frame

∽

III. NAVIGATE

1. Prepare to begin.
2. Set *'Time Flow'* to be *'continuous'* — make certain it does <u>*not*</u> (yet) jump from one 'jump' to the next.
3. Set flow rate to be something like *6000 x* (in comparison to 'real time').
4. *'Play'* a *'**Day**'*: the interval between one rising of the sun and then next.

~

*P*lay another.

~

*P*lay another,

but this time a little slower (e.g.: *3000x*).

~

*P*lay another,

but this time for two *Day*s, etc.

~

*P*lay for approx. six days. Rest on the seventh.

Get ready: it's time to take the world out for a spin.

~

*Z*oom in. Zoom out.

Go somewhere. Go somewhere else.

Look. See.

~

*S*ee it as it would have been seen January 2007.

See it as it will have been seen January 2019.

~

*S*eek. Hide.

Lose. Find.

SATELLITES BY JOVE

꧁

2007. JULY 26

This week, Jupiter continues to occupy the celestial throne. Around 9:30 pm, face almost due south and tilt your head up just a bit. If the clouds cooperate, you cannot miss the gas goliath. It will be by far the brightest object in its neighborhood (other than the moon). Jupiter has a slightly yellow hue, but intensity alone will distinguish it.

If you happen to have even a low-grade pair of binoculars, check out the line of four dots that pop into view when you take a closer look at Jupiter. These are moons of Jupiter: *Io*, *Europa*, *Ganymede* and *Callisto*. An ocean surface and oxygen atmosphere (among other features) make Europa an especially attractive candidate for the possibility of extra-terrestrial life.

A number of the stars near Jupiter will be obscured by the moon's abundant light. As of Thursday, July 26, the moon's shape will be *gibbous* (three-quarters of a circle). It will continue to grow until reaching full phase on Sunday, July 29. On that day, the moon will rise in the east at approximately the same time (8:20 pm) that the sun will set in the west. Each night, as the moon rises earlier by almost an hour, you will find it farther and farther east (left) of Jupiter.

After Sunday, you will begin to notice a shadow creep over the moon's right side; it will continue to wane for two more weeks.

Planet sightings rarely fail to disappoint. If you wish to view a planet without the distraction of lunar light, then treat yourself to an early morning stroll. The sun rises in the east, this week at approximately 5:40 am. During late July, each day break will be presaged by the sun's messenger: Mercury. This rocky, crater-filled neighbor of ours is diminutive (more than 15,000 times smaller than Jupiter) and closely tethered to the sun so it is frequently upstaged.

Right now, however, is a great time to meet Mercury. It spends the end of July keeping a particularly comfortable distance from both the moon and blazing sun. Watch the eastern horizon at around 5:10 am. That solitary bright 'morning star', rising valiantly until its inevitable wash-out, is the planet Mercury.

August 2007 will continue to reward early-morning observers. The *Perseid* Meteor Shower (peak: August 13) and a total lunar eclipse (August 28) offer two spectacular reasons to jump out of bed.

SPICA BY SATELLITE

❦

AUGUST 16

*T*his past Monday, August 13, the moon was new. If you were to imagine a 93 million mile line connecting the Earth to the sun, then the moon was on that line — in between the two other objects. Any sunlight that hit the moon, therefore, bounced right back to the sun and never made its way to our planet. Since the moon does not produce its own light, it was not visible to us.

Now the moon continues its monthly sweep away from the sunny side to the outer or 'night' side of Earth. As it gets closer and closer to the night side of Earth, more and more of the sun's rays can both reach the moon and reflect toward us. An increasingly greater portion of the moon's disc is visible and the moon appears to 'wax'.

This week, the moon will wax from a slim crescent to a hearty *gibbous* (three- quarter). If you look to the low west shortly after sunset (approx. 7:53 pm) the night of Friday, August 17, you'll find that the crescent has some company. Directly to the left of the moon and toward its top will be a singularly vivid star: the most obvious object in the moon's neighborhood. This star is *Spica*.

Spica is the brightest star in the zodiac constellation *Virgo*. It is the 15th brightest of *all* stars visible from anywhere, anytime, on Earth. Spica is approximately one and half times as massive as our sun. It lives approximately one and a half *quaδrillion* miles away from us. This means that for a bit of Spica's light to reach your eye, it has to make a journey that lasts approximately 260 years. The Spica you see is actually the Spica that existed before the American Revolution. What of the Spica that exists now? We have no way of knowing if there even is one.

We nonetheless expect Spica to remain home in Virgo each night after August 17. But the moon will continue to slide to the left—farther and farther away from the constellation. Spica will thus become more and more difficult to identify.

Half of the moon will be illuminated Monday August 20. One would expect the moon to appear full roughly a week later—on Tuesday, August 28. This month, however, things will be delightfully different. Nature has lined things up for the totality of the moon to be *eclipsed* by Earth's shadow! Stay tuned.

TALE OF TWO MOONS: TOO TALL, ALL TOLD

On Tuesday morning, August 28, mother nature will provide the final answer to a question sent in by one of our star-gazing readers. The question is:

> Will I really be able to see *two* moons in the sky next week?"

The question may seem strange and out of nowhere, but it is, in fact, only strange.

Eager anticipation of 'two moons' has been generated by a number of recent articles and emails that have been blazing trails around the web—particularly among astrology communities. The emails forecast a 'once-in-a-lifetime' close approach of the planet Mars. According to the forecast, Mars will be so unusually close to Earth, and so nicely positioned relative to the sun, that it will look as big as our moon to the naked eye. If you look up at the right time, concludes the argument, you will enjoy a moment previously available only to Luke Skywalker: During the unfolding of the 1973 Star Wars film, Luke contemplated the meaning of a mundane farm life while two suns set in the background of his home planet *Tatooine*.

The rumor's logic itself is not outlandish. Mars's distance from Earth varies significantly. When Mars is slightly closer, it does appear slightly larger. If Mars happens to be on Earth's 'night-side' (opposite from the sun) during a close approach, it can look unusually large and bright. The orbit of Mars is, moreover, quite predictable.

BUT: This is where the 'two moon' prophecy runs into serious trouble. Mars will not even come close to close enough for the presentation of such a prodigious disc. It never has. The disappointing reality is that the 'two moon' emails have been recycled and re-circulated every mid-summer since 2003. Back then, Mars's proximity was indeed record-setting. It was quite gratifying to see in a telescope, but Mars, even that summer, struck the naked eye as a nice, bright, orange-hued star.

Mars is approximately twice as large as our moon. In order for it to appear as large as our moon, therefore, it would have to sit no more than twice as far away.

Currently, Mars is approximately 116 million miles away, while the moon is only 226 **thousand** miles. So Mars is more than 500 times farther from us than the moon. Monday night's full moon, alas, will fly definitively solo.

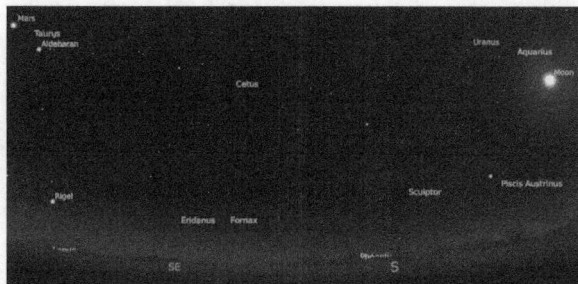

Nonetheless, if you watch the full moon setting in the west around 4:45 in the morning of Tuesday, August 28, you will certainly see something worth seeing. The moon will be directly on the Earth's night side (hence, 'full'); more remarkably, it will gradually fall smack onto the Earth-sun line and into Earth's shadow: It will undergo a total eclipse. Instead of 'two moons', we will experience the vanishing of one.

Total lunar eclipses hardly occur every month, because the moon is usually somewhat above or below the Earth-sun line. To watch a shadow creep over and eventually cloak the entire moon is sublime. From our New England longitudes, however, this will all be occurring both while the moon sets and while, in the east— following the rising of a star-sized Mars—the sun rises (each at approx. 6:15 am). So you will want a really flat landscape and strong cup of coffee to catch the act before the occulted moon drops below a brightening horizon. My pot is brewing.

CASSIOPEIA

༄

The southern night skies should be nice and tranquil this week. Mercury and Jupiter retire early in the evening; Venus, Saturn and Mars hide until dawn; the moon is on the wane and thus rises at increasingly wee hours. On Friday, September 1, it will not even have risen until 9:00 pm. The highest, most visible, point in the moon's climb will not be reached, therefore until roughly 3:00 am on the 2nd — and then later and later by an increment of approximately 40 minutes each subsequent night. Within a week, the moon's path will have fully become a morning phenomenon. This is utterly

typical and, in fact, analytically demon-
strable for any waning moon. Simply but
perhaps surprisingly put, the moon is not
'out' in the night any more than it is during
the day: It's just easier to see.

To many a casual up-looker, the sight of a
morning moon can seem quite surprising.
And forget about that one crazy time
they'll swear to have seen — all sober and
sentient — some crazy audacious last
quarter moon who lingered flagrantly at
the center of sky stage and dared to duet
during sol's time to solo. A sun-moon
pairing is utterly routine — as regular as
rent — but no need to rain on anyone's
parade.

The northern sky beckons. Pick any clear
night and notice where the sun sets. This is
west. Once dark has fallen, rotate your
body ninety degrees clockwise. You are
now facing north. If you look somewhat
low in the sky, you should have little
trouble spotting the *Big Dipper*. This group
of seven stars is unusually large; it earns its
name by resembling an enormous soup

ladle, with the bowl down and open to the right.

Look at the handle of the Big Dipper and find what appears to be the second star from the handle's edge. This area of the dipper is comparatively bright because there is actually more than one star bunched together—one in back of another. Although the star known as *Alcor* is roughly 15 trillion miles behind the star called *Mizar*, our eyes are too far away from either to make the distinction. We just end up with a hearty gob of light.

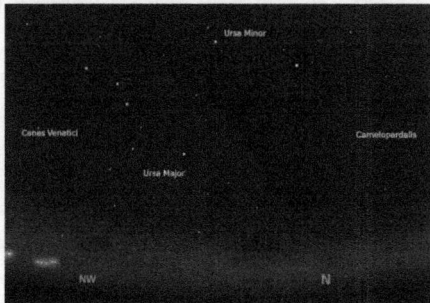

Starting from Mizar's home in the Big Dipper, draw an imaginary line east and slightly up. If you follow this line the rough equivalent of two handle lengths, you will hit the end of another, smaller, spoon. You have reached *Polaris*: the edge of the *Little Dipper*. (The Little Dipper can be difficult to identify on its own, but it pops nicely into view when the Big Dipper is used as a guide.)

Once you have hit the end of the Little Dipper, just keep following your line.

Trace it the same distance that you traveled from one Dipper to the next. You should find yourself in the midst of a hefty but angular 'three' consisting of five stars. Depending on the time of night or the time of year, this 'three' can look like the letter *E*, the letter *M* or the letter *W*. It rotates continuously around Polaris.

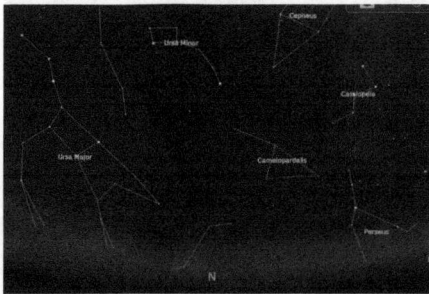

This prominent constellation is *Cassiopeia*.

Its most dazzling star is *Schedar*, the second one from the top. In Greek mythology, the magnificent Queen Cassiopeia, mother to Princess Andromeda, was a bit too aware of her own beauty. Punishment for her vanity almost led to the sacrifice of Andromeda. Luckily, Andromeda was saved.

The immortality of Cassiopeia's daughter was preserved in the name of our nearest-neighboring galaxy: the *Andromeda* galaxy. On extremely clear nights, this spiral star world, approximately 2.5 million *light-years* away, can actually be observed without a telescope. The associated mythology, perhaps, helps add a bit of color or flavor to whatever runs through the mind while attempting to consider the sight of another galaxy. The purely scientific implications of a look at Andromeda, however, ought suffice to stagger.

Consider first that something one light-year away is something approximately 6 trillion miles away. And you're looking at it. Consider and/or recall second that for you to see something — anything — light simply must travel to your eye from the thing. How else? In this case, then, even the faintest blur of illumination you glimpse is entirely and only what was there to see when the light left: which is to say approximately, no backing out of the math now, 2.5 million years ago.

So, just look to the right of Schedar, roughly a full Cassiopeia length. Squint if you have to: To gaze at the Andromeda Galaxy is literally to look as far out in space — and therefore as far back in time — as the (naked) eye can see.